RED BUD FARM®

Color and Learn With Buddy the Worm!

Let's get dirty!

Buddy

RED BUD FARM
Growing Green!

Remember, all worm castings are NOT the same!

RED BUD FARM worm castings  OMRI® LISTED — For Organic Use

are OMRI (Organic Materials Review Institute) listed.

This book is designed to be educational for children of all ages, as they are instructed while they color about the many benefits of organic **RED BUD FARM** worm castings. Products of **RED BUD FARM** are not available for purchase outside the USA.

RED BUD FARM

PO Box 770
Blountville, TN 37617-0770
www.red-bud-farm.com
email red_bud_farm@aol.com

Buddy is the **RED BUD FARM** worm!

Buddy wants to teach you important things about how worm castings help you grow all kinds of plants. When you plant with worm castings, you don't need anything else! Just mix them in the dirt outside or the potting soil inside and you're good to go! So color and learn with Buddy the Worm!

What are Red Bud Farm worm castings?

'Castings' are what worms 'cast off'; they are worm manure. The fancy name is 'vermicast'. Earthworms are some of the cleanest, most helpful creatures in the world! As Buddy and his friends move through the dirt, they make space for air and water so plant roots can grow bigger and stronger. Worms are the animals that are in the dirt to make things grow.

Red Bud Farm
worm castings are child-safe and pet-friendly.

Kids can play with worm castings just like they do in dirt!
Children can touch or handle worm castings without gloves.
Castings feel clean and brush right off your hands.
Animals will not be hurt by worm castings either.
So if dogs or cats paw or scratch in them, they will not get sick.

Castings help inside and outside plants.

Buddy

Red Bud Farm worm castings fertilize plants.

Fertilizer is a substance that is added to potting soil indoors or to dirt outdoors to help all kinds of plants grow. Using Red Bud Farm worm castings makes plants fertile, which means they grow easier, faster, bigger, and better than ever! Organic worm castings are all-natural fertilizer. They have no chemicals.

When should you fertilize plants?

1 - Fertilize seeds and plants when you plant them in the spring or fall. When you dig a hole, drop a big handful (about 1 cup) of Red Bud Farm worm castings right there at the root level. Mix another heap of castings with the dirt that you pulled out, and put it back in around the roots for the backfill.

2 - Use worm castings as a top or side dressing like a fertilizer. Scratch around the stem to loosen the soil, and pack a couple of handfuls of castings there. Rain and watering will take them down in the ground to give the roots a quick boost of nutrition.

3 - Fertilize anytime plants need water by making a worm TEA. Pour it directly, or strain it into a pump sprayer and spray it.

Red Bud Farm worm castings will not burn.

Chemical fertilizers can harm plants and need to be measured. Too much chemical fertilizer will cause root damage that can make leaves wilt, turn brown, and die. Unlike chemicals, Red Bud Farm worm castings will not damage plants, even if they are not watered immediately. They have no chemicals, no artificial ingredients, and no man-made stuff in them at all.

Red Bud Farm worm castings are the best animal manure.

Other fresh, raw manures (like chicken, bird, bat, cow, horse, sheep, goat, rabbit, etc.) can take months to break down and compost. But worm castings are ready to use immediately and directly on all plants, even when starting seeds.

NEVER put pig, dog, cat, or human waste on plants because they might contain germs or diseases that can infect people.

Red Bud Farm
worm castings are the best plant food.

They work fast and you can see quick results in just a few days. They also work slowly as they supply nutrients up to a year or more. That is why we say they are 'time-released'. Castings have a highly active mixture of helpful enzymes and minerals necessary for excellent plant growth. They fight off diseases so toxins, poisons, and destructive bacteria will not harm plants.

Red Bud Farm worm castings have no odor.

Many other kinds of manure smell bad, but not worm castings!
Even when you smell them .... you can't smell them! SWEET!
Worm castings look and feel like very dark-colored coffee
grounds. They aren't sticky, and will brush right off your hands.

Worm castings are good for the environment.

Worm castings do not pollute the ground. They are safe for the
entire eco-system and reduce algae in ponds and greenhouses.

Red Bud Farm worm castings make room for air.

Roots need oxygen to grow and spread out in the ground. Worm castings act like little blocks of wood in a bucket to hold space open in the dirt so plants can breathe and expand.

Worm castings hold moisture, but not too much.

In the soil, worm castings act like tiny sponges that hold water for plants to drink whenever they are thirsty. Storing water in soil makes it slightly damp so roots do not get too dry. Worm castings also soften the dirt so roots can grow through it. Then excess water can drain off to protect plants from root rot.

Red Bud Farm worm castings help repel insects.

Worm castings are known to run off pesky spider mites, white flies, aphids, many harmful bugs, and parasites that feed on plant juices. Worm tea can be sprayed on leaves to help also.

Red Bud Farm worm castings help fight off lots of plants' sicknesses, diseases, and viruses.

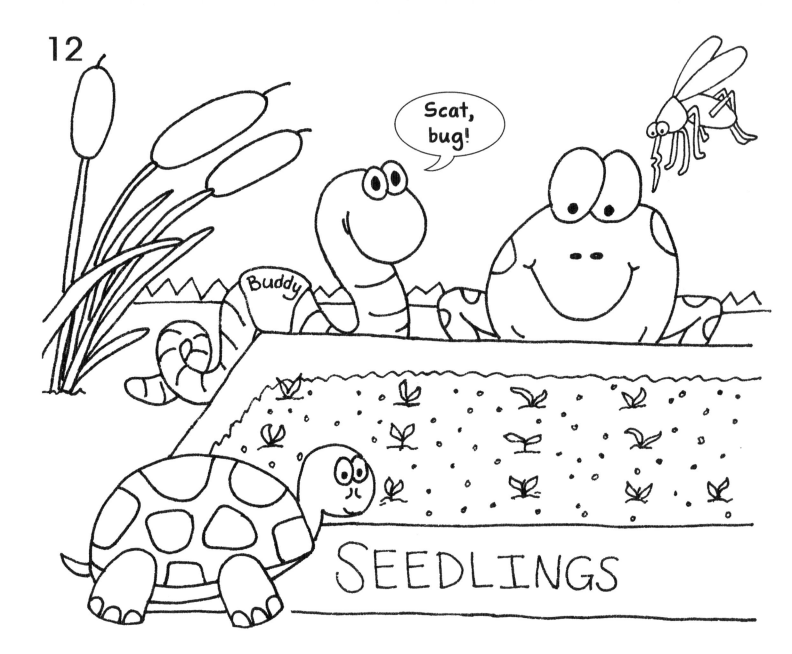

Start seeds with Red Bud Farm worm castings.

Make a mixture of 1 part worm castings and 3 - 4 parts potting soil, and put it in small containers (such as egg cartons, small plastic cups, or flats). Cut holes in the bottoms so water can drain. Space out and press seedlings about 1/4" under the soil or follow seed package instructions. Water to keep them moist. Sit them under a fluorescent light or window to help sprouting.

When seedlings are 4 - 6 inches tall, move them outside for a few hours each day for a week so they won't be shocked by going outdoors. Then transplant them into a garden with worm castings in warm weather after the danger of frost is gone.

Grandparents give seeds to parents, who save seeds for their children to plant and harvest, to pass them along to their kids.

Grow food with Red Bud Farm worm castings.

Seeds that have been passed down at least 60 years from generation to generation are called 'heirloom' seeds because you inherit them. They reproduce the same quality every year. They are non-hybrid and open-pollinated. They have not been treated or sprayed with chemicals or pesticides like GMO seeds (Genetically Modified Organism) that are changed in the seed.

Red Bud Farm worm castings can be stored.

A greenhouse, garage, storage shed, or basement are all good places to keep bags of worm castings until you are ready to put them in your plants. They will still be as potent (powerful) when you use them as they were when you first got them. They don't go bad in any temperature. Store them in a dry place.

Red Bud Farm worm castings work a long time.

They supply nourishment like time-release vitamins to plants to keep them healthy and strong all through the growing season. They last for days, weeks, months, and even years.

Red Bud Farm Worm Castings work everywhere.

You can plant with Red Bud Farm organic worm castings ....
* indoors in pots and seedling trays
* in greenhouses
* outside in the dirt
* in raised garden beds
* in containers or buckets   .... everywhere plants grow!

Use Red Bud Farm worm castings indoors.

Put worm castings in the pots around all your inside plants to revive old potting soil and bring forth new buds and leaves. All kinds of colorful flowers and beautiful green potted plants will grow like never before. Worm castings help keep away lots of different kinds of insects that cause trouble.

Indoor plants exchange the air in your home because they take in a gas called carbon dioxide (that people exhale) and they put out oxygen for you to breathe in. So plants inside your house make the air fresher and cleaner. They are natural air purifiers .... healthier than candles, fragrance sprays or incense!

Use Red Bud Farm worm castings outdoors.

Put castings in the holes and around all your outside plants, too.
*Plant a garden of fruits and vegetables with worm castings.
*Add them in flower beds when and after you plant with
    annuals (flowers that only bloom the year you plant them)
    and perennials (flowers that bloom every year).
*Loosen the dirt around every shrub or bush and pack a couple
    of handfuls of worm castings at the stem base.
*Make worm tea to pour around tree trunks and landscaping.
*Fertilize the entire yard by spreading worm castings all over
    grass, or scratch up bare spots and add them with seed.

Farmers love Red Bud Farm worm castings.

Homesteading is about being self-sufficient. Many country folk enjoy growing their own food. They like to know that what they eat is wholesome and natural.   Some sell it at farmers markets.

City dwellers and urban farmers use Red Bud Farm worm castings.

Folks who do not own land, or live in condominiums, or rent apartments, use worm castings to grow plants, flowers, fruits, vegetables, and herbs. They plant in containers, on porches, individual pots, sunrooms, window sills, or raised beds in yards.

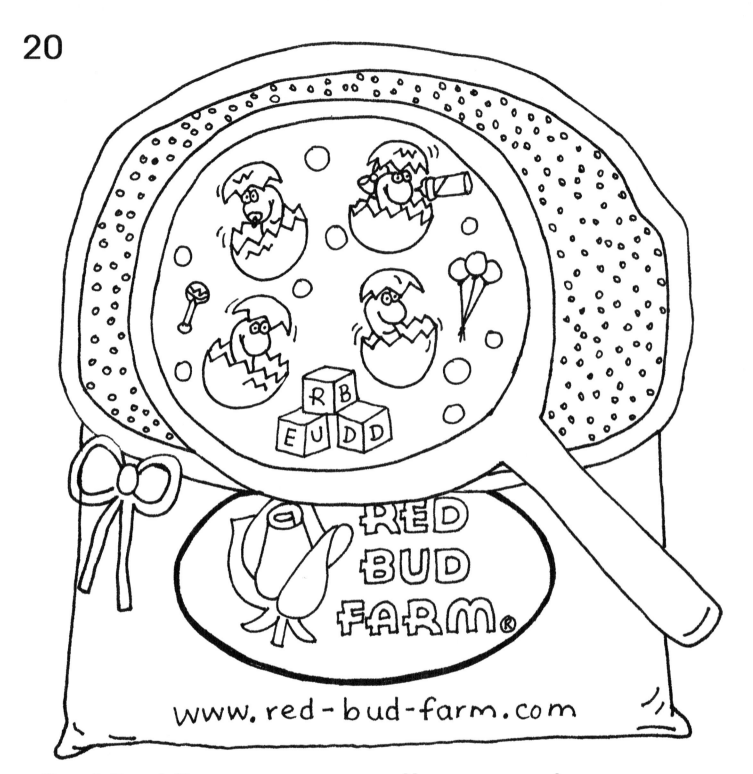

Red Bud Farm worm castings may have eggs.

During the process of separating the worms from their castings, sometimes tiny sacks of baby worms can filter through and get packaged inside the bags. These little brownish-yellow eggs are called 'cocoons' and are shaped like lemon drops or balls. If they hatch, there can be 1 to 7 itty-bitty worms come out. When these babies are born, they can grow into little Buddies!

Red Bud Farm Worm Castings Tea Recipe:

Follow these easy steps to make worm tea for plants.
Use it to water or spray plants for a quick boost of health.

1. Start with a 5 gallon bucket of water.
   - If you use well water or rain water, go to step 2.
   - If you pay a water bill, then let your water sit for 24 hours
     so the chlorine will dissolve out of it.
2. Stir in 1 quart of Red Bud Farm worm castings into the water
   with the end of a broom or a long stick.
3. For 24 hours, stir the tea in the bucket several times
   ---OR--- hook up an aquarium bubble stone to 'brew' it.
4. POUR the tea on plants to water them.
5. OPTIONAL: You may want to SPRAY the tea on plants. If so,
   ADD 2 tablespoons of unsulphured molasses (or corn syrup)
   to make it tacky enough to stick on plant leaves, and STIR.
   Then STRAIN the tea with nylons or layers of cheesecloth
   (so the tea will not clog the nozzle) as you pour it into
   a pump sprayer and SPRAY it all over your plants.
6. Some of the worm castings that did not dissolve will still be
   left in the bucket. They are still good and active, so apply
   them around the base of any plants to help them grow.
   Use all the tea within 24 to 36 hours.

CPSIA information can be obtained
at www.ICGtesting.com
Printed in the USA
LVOW05s1332081115

461107LV00001BA/1/P